The Dodo's

The Fascinating Journey of the Dodo

Copyright © 2023 by Ravi Ramdenee-Soobhug.

All rights reserved. No part of this publication may be reproduced, stored in a retrieval system, or transmitted, in any form or by any means, electronic, mechanical, photocopying, recording, or otherwise, without the prior written permission of the copyright holder. This book was created with the help of Artificial Intelligence technology.

The contents of this book are intended for entertainment purposes only. While every effort has been made to ensure the accuracy and reliability of the information presented, the author and publisher make no warranties or representations as to the accuracy, completeness, or suitability of the information contained herein. The information presented in this book is not intended as a substitute for professional advice, and readers should consult with qualified professionals in the relevant fields for specific advice.

A Mysterious Island Discovery 6

The Wondrous World of Mauritius 9

Unveiling the Dodo's Origins 12

The Dodo's Taxonomic Journey 15

The Dodo's Physical Characteristics 17

A Flightless Wonder of the Avian World 20

Adaptations for Island Life 23

The Dodo's Diet and Feeding Habits 26

The Dodo's Social Behavior 29

Encounter with the First Humans 32

Early Observations and Specimens 34

Naturalists and the Dodo's Fame 37

The Golden Age of Exploration 40

Ecological Impact of Human Presence 43

Hunting and Predation on Dodos 46

Human Exploitation and Extinction 49

The Last of the Dodos 52

Legends and Myths Surrounding the Dodo 55

Rediscovering the Dodo in Art and Literature 58

Scientific Analysis and Reconstruction 61

Clues from Fossil and Subfossil Evidence 64

The Dodo's Place in Evolutionary History 67

Conservation Efforts and Lessons Learned 70

Modern-Day Representations of the Dodo 73

The Legacy of the Dodo: Lessons for Future Generations 76

A Mysterious Island Discovery

In the vast expanse of the Indian Ocean, there lies a small tropical island known as Mauritius. Its story begins centuries ago when intrepid sailors embarked on daring voyages in search of new lands and untold treasures. It was during one of these seafaring adventures that the intriguing tale of the dodo begins.

Picture a scene filled with anticipation and wonder as Portuguese explorers, sailing under the command of Captain João da Nova, stumble upon the enchanting shores of Mauritius in the early 16th century. The island, veiled in dense forests and lush vegetation, seemed like an undiscovered paradise waiting to be explored.

As the crew of the ship ventured inland, they encountered a sight unlike any they had ever witnessed before. It was a peculiar bird, plump and flightless, with a distinct appearance that set it apart from all other avian species. This remarkable creature would later be named the dodo, a name derived from the Portuguese word "doudo," meaning "foolish" or "simpleton."

The dodo stood about a meter tall, with a rounded body covered in grayish-brown feathers. Its wings were small and feeble, rendering it unable to take flight. The bird's most distinguishing feature was its large, hooked beak, which was perfectly suited for its diet and foraging habits. With short, stout legs and a waddling gait, the dodo moved with a certain charm and grace, despite its clumsy appearance.

Word of this fascinating avian discovery quickly spread throughout seafaring circles, capturing the imagination of naturalists, scientists, and adventurers alike. The dodo's unique characteristics and its apparent lack of fear towards humans sparked curiosity and intrigue. Many were eager to learn more about this newfound species and unravel the mysteries surrounding its existence.

Unfortunately, as the years passed, the dodo's fate took a tragic turn. The arrival of humans on Mauritius brought with it unforeseen consequences for this remarkable bird. Sailors and explorers, hungry for fresh meat during their long sea voyages, found the dodo to be an easy target. Its trusting nature, lack of fear, and limited mobility made it vulnerable to the relentless hunting pressure exerted upon it.

As human settlements took root on the island, the dodo faced additional threats. The introduction of invasive species, such as rats, pigs, and monkeys, disrupted the delicate balance of the island's ecosystem. These new arrivals competed with the dodo for food and destroyed its natural habitat, further pushing the species towards the brink of extinction.

By the late 17th century, just over a century after its initial discovery, the dodo had vanished from the face of the Earth forever. The last remnants of this once-majestic bird were reduced to bones and fragments, preserved in the annals of history and the collections of museums worldwide.

The dodo's extinction serves as a sobering reminder of the impact humans can have on fragile ecosystems and the need for conservation efforts to protect the world's biodiversity. Today, Mauritius and its surrounding islands strive to preserve their unique flora and fauna, learning from the mistakes of the past.

Although the dodo may no longer grace the forests of Mauritius with its presence, its legacy lives on. The bird's fascinating story has captivated the minds of scientists, artists, and nature enthusiasts, inspiring countless works of literature, art, and even symbolizing the fragility of life itself.

In the chapters that follow, we will delve deeper into the history, biology, and cultural significance of the dodo. We will explore its origins, its adaptations to the island environment, and the unfortunate events that led to its demise. We will also examine the lasting impact of the dodo on our understanding of extinction, conservation, and the interconnectedness of all living beings.

So, join me on this remarkable journey as we uncover the mysteries and unravel the history of the dodo, a bird that once roamed the enchanting shores of Mauritius, forever etched in the annals of time as a symbol of both wonder and warning.

The Wondrous World of Mauritius

Nestled in the azure waters of the Indian Ocean, Mauritius is a paradise that exudes natural beauty and captivating landscapes. This enchanting island, located approximately 2,000 kilometers off the southeastern coast of Africa, boasts a rich and diverse ecosystem that has fascinated explorers and naturalists for centuries.

As we step into this wondrous world, we are greeted by an abundance of breathtaking sights. The island's coastline stretches for more than 330 kilometers, adorned with pristine beaches of powdery white sand and fringed by vibrant coral reefs teeming with marine life. Crystal-clear turquoise waters lap gently against the shore, inviting us to immerse ourselves in the wonders that lie beneath the surface.

Beyond the shoreline, Mauritius reveals a captivating tapestry of landscapes. Verdant rolling hills carpeted with emerald-green sugarcane fields create a picturesque patchwork that stretches as far as the eye can see. Towering mountains and rugged cliffs provide a dramatic backdrop, showcasing the island's volcanic origins.

Venturing into the heart of Mauritius, we discover an extraordinary wealth of biodiversity. The island's tropical rainforests are a treasure trove of endemic flora and fauna, found nowhere else on Earth. Dense canopies of ancient ebony trees, with their dark, lustrous wood, create a sense of enchantment as dappled sunlight filters through the foliage.

The vibrant birdlife of Mauritius is a testament to its ecological richness. The rare echo of the Mauritius kestrel's call can be heard echoing through the forest, a success story of conservation efforts that saved this once critically endangered species from the brink of extinction. The colorful flashes of the Mauritius parakeet and the playful antics of the Mauritius cuckoo-shrike add to the symphony of avian life that fills the air.

But perhaps the most famous and enigmatic resident of Mauritius was the dodo. This flightless bird, now extinct, once roamed the island's forests and coastal regions. Its presence was a testament to the unique evolutionary journey that unfolded in isolation on this remote island. The dodo was not alone in its isolation; other endemic creatures, such as the giant tortoise and the pink pigeon, thrived in Mauritius' pristine habitats.

As we delve deeper into the island's ecosystems, we discover hidden gems of natural wonders. Hidden cascades of water plunge into crystal-clear pools, forming tranquil oases amidst the lush foliage. Unique plant species, such as the rare Mauritius ebony and the delicate orchids, paint the landscape with bursts of color and fragrance.

The coastal waters surrounding Mauritius are a playground for a variety of marine life. Coral reefs, with their kaleidoscope of colors, shelter a plethora of fish, from vibrant angelfish to graceful sea turtles. Dolphins dance in the waves, their playful antics a joy to behold, while majestic whales make their annual migration, breaching the surface with grace and power.

Mauritius is not only a haven for flora and fauna but also a melting pot of cultures and traditions. Its history is woven with the narratives of Dutch, French, and British colonialism, each leaving their mark on the island's architecture, cuisine, and way of life. The vibrant mix of Creole, Indian, Chinese, and European influences creates a unique cultural tapestry that enchants visitors from around the world.

As we conclude our exploration of the wondrous world of Mauritius, we are left in awe of its natural splendor and the delicate balance that sustains its ecosystems. It is a place where nature thrives, and the human spirit finds solace in the embrace of its beauty.

Unveiling the Dodo's Origins

In our quest to understand the dodo, we embark on a journey through time to unveil the origins of this fascinating bird. Tracing its lineage and unraveling the mysteries of its evolutionary history brings us closer to comprehending the unique creature that once inhabited the island of Mauritius.

To begin our exploration, we must delve into the world of avian evolution. Birds, as we know them today, are descendants of a group of theropod dinosaurs that roamed the Earth millions of years ago. Through a gradual process of adaptation and diversification, these ancient dinosaurs developed feathers and took to the skies, eventually giving rise to the incredible diversity of bird species we observe today.

The dodo's closest relatives are believed to be pigeons and doves, belonging to the family Columbidae. This family encompasses a wide array of bird species found across the globe, ranging from the small and delicate diamond doves to the majestic crowned pigeons of New Guinea. Through comparative analysis of anatomical features and genetic studies, scientists have established the dodo's affiliation with this avian family.

The ancestral lineage that led to the dodo's evolution likely originated in the vicinity of Southeast Asia. It is believed that some ancestral pigeons dispersed across the vast expanse of the Indian Ocean, colonizing various islands along their journey. One such island was Mauritius, where the dodo's story would eventually unfold.

The isolation of Mauritius played a pivotal role in shaping the evolutionary trajectory of the dodo. With no land predators to contend with and an abundance of food sources available, the ancestral pigeons that arrived on the island found themselves in a unique ecological niche. Over time, these birds underwent adaptive changes to better suit their newfound environment, giving rise to the flightless wonder we now know as the dodo.

The dodo's flightlessness is a remarkable adaptation that allowed it to conserve energy and explore the terrestrial habitats of Mauritius. With no need to take to the skies, the dodo's wings gradually reduced in size and became less functional. This transformation freed up energy and resources that could be channeled into other aspects of its biology, such as its distinctive beak and robust physique.

The dodo's beak, with its characteristic hook and stout structure, played a vital role in its foraging habits. It allowed the bird to exploit the resources available on the island, such as fallen fruits, seeds, and nuts. The dodo's diet was varied and included both plant matter and the occasional insect or small animal. Its beak was a versatile tool, capable of cracking open tough shells and extracting the nourishment within.

While the dodo's evolutionary journey on Mauritius was shaped by its flightlessness and specialized feeding habits, it also underwent changes in its size and appearance. Fossil evidence suggests that earlier populations of dodos were smaller in size compared to their later counterparts. This phenomenon, known as insular dwarfism, is often observed when species are isolated on islands with limited resources. Over time, the dodo's size increased, likely due to the absence of predation and the abundance of food sources on the island.

The dodo's physical characteristics, such as its plump body, short legs, and waddling gait, were adaptations that enabled it to navigate its terrestrial environment. While these traits may have appeared comical or clumsy, they were perfectly suited to the dodo's way of life. Its rounded body provided insulation and storage for energy reserves, while its sturdy legs allowed it to traverse the forest floor with relative ease.

As we unravel the origins of the dodo, we come to appreciate the intricate web of evolutionary processes that brought this remarkable bird into existence. The combination of flightlessness, specialized beak morphology, and adaptations to its island habitat allowed the dodo to thrive in Mauritius for thousands of years.

The Dodo's Taxonomic Journey

In our exploration of the dodo, we now turn our attention to its taxonomic journey—the process by which scientists classified and categorized this remarkable bird. Taxonomy, the science of naming and classifying organisms, plays a crucial role in understanding the relationships between species and their place in the tree of life.

The dodo presented taxonomists with an intriguing puzzle. When it was first discovered by European explorers in the 16th century, it was unlike any bird they had encountered before. Its unique characteristics and endemic nature sparked debates and discussions among naturalists who sought to understand its place within the larger avian family.

Early descriptions of the dodo's physical attributes, combined with its inability to fly, led some naturalists to compare it to the ostrich, a large flightless bird found in Africa. However, closer examination and scientific scrutiny revealed distinct differences between the two species. The dodo possessed a more robust build, a distinctive beak shape, and a unique combination of anatomical features that set it apart from the ostrich.

As more information became available, taxonomists began to recognize the dodo as a member of the family Columbidae, which includes pigeons and doves. Comparative anatomy and genetic studies provided evidence for this classification, revealing similarities in skeletal structures, internal anatomy, and DNA sequences.

Within the family Columbidae, the dodo was assigned its own genus, Raphus, indicating its uniqueness among its pigeon relatives. The scientific name given to the dodo, Raphus cucullatus, reflects both its physical attributes and its resemblance to other pigeons within the genus.

However, the taxonomic journey of the dodo did not end there. Over the years, as scientific knowledge advanced and new techniques were developed, taxonomists refined their understanding of the dodo's classification. Further analysis of skeletal remains and genetic material allowed for a more precise assessment of its relationships to other birds.

Today, the dodo is recognized as part of a larger group called the Raphinae, which includes not only the dodo but also its closest relative, the Rodrigues solitaire. These two birds, both flightless and endemic to islands in the Indian Ocean, share common ancestry and are considered part of a unique evolutionary lineage within the pigeon family.

The taxonomic journey of the dodo serves as a testament to the dynamic nature of scientific inquiry. As new evidence emerges and techniques evolve, our understanding of species relationships and classifications continues to evolve as well. The dodo, once a subject of curiosity and speculation, has found its place in the intricate web of avian taxonomy, connecting us to its lineage and highlighting the interconnectedness of all life forms.

The Dodo's Physical Characteristics

The dodo, with its unique appearance and charm, captured the imagination of those who encountered it during its brief time on Earth. Let us delve into the fascinating details of this remarkable bird's physical traits.

Standing at approximately one meter tall, the dodo was a relatively large bird. Its rounded body, covered in grayish-brown feathers, gave it a distinctive and somewhat plump appearance. The feathers provided insulation and protection from the elements, allowing the dodo to thrive in the diverse habitats of Mauritius.

One of the dodo's most distinguishing features was its head, with its remarkable beak. The beak of the dodo was large, stout, and slightly curved, showcasing its adaptation to a specific diet and foraging behavior. This specialized beak allowed the dodo to efficiently extract food from its environment.

The dodo's beak was perfectly suited for its herbivorous diet. It had a strong and hooked structure, which enabled it to crack open the tough outer shells of fruits and nuts. This remarkable tool provided access to the nutritious contents hidden within, allowing the dodo to exploit a wide range of food sources found on the island.

Beneath its beak, the dodo sported a small, dark-colored eye. While the exact function and visual capabilities of the dodo's eye remain a subject of scientific inquiry, it is believed that the bird had relatively good eyesight, which would have aided in navigating its surroundings and identifying potential food sources.

The dodo's wings, in contrast to its robust body, were small and feeble. These wings, though reduced in size, still retained some feathers and skeletal structures associated with flight. However, the dodo's wings were not functional for sustained flight, rendering the bird flightless. Instead, the dodo relied on its strong legs and ability to navigate the forest floor.

Speaking of legs, the dodo had short and stout legs, which were well adapted for its terrestrial lifestyle. The sturdy legs and feet allowed the dodo to traverse the varied landscapes of Mauritius, from the dense forests to the coastal regions, with relative ease. Its gait, often described as a waddle, added to its unique charm and set it apart from other birds.

When it comes to the dodo's coloration, historical accounts and artistic representations suggest that it had predominantly grayish-brown plumage. However, it is important to note that the accuracy of these depictions can be influenced by factors such as artistic interpretation and the passage of time. Further research and analysis of preserved specimens provide valuable insights into the dodo's coloration and pigmentation.

The dodo's appearance and physical characteristics were a testament to its unique evolutionary journey on the isolated island of Mauritius. Over time, the bird underwent adaptations that allowed it to thrive in its environment, shaping its morphology and behavior in response to the challenges and opportunities presented by its island home.

As we unravel the physical characteristics of the dodo, we gain a deeper appreciation for the intricacies of this remarkable bird. Its rounded body, distinctive beak, flightless wings, sturdy legs, and overall charm make it a truly captivating subject of study and exploration.

A Flightless Wonder of the Avian World

In the vast realm of avian wonders, the dodo stands as a remarkable testament to the incredible diversity and adaptability of birds. Among the myriad species that grace our skies, the dodo holds a special place—a flightless wonder that captivates our imagination and challenges our understanding of avian evolution.

Flight has long been considered a defining characteristic of birds, allowing them to traverse vast distances, conquer different habitats, and access a multitude of resources. However, the dodo emerged as a unique exception to this rule. Unlike its airborne counterparts, the dodo was flightless, grounded by the demands of its environment and the evolutionary forces that shaped its existence.

The loss of flight in the dodo was not an overnight transformation but a gradual process that unfolded over generations. The absence of land predators on the isolated island of Mauritius provided a unique opportunity for ancestral pigeons to explore new ecological niches, free from the need for aerial locomotion. Over time, these birds adapted to their island home, trading flight for other advantageous traits.

The dodo's wings, though reduced in size, still retained some feathering and skeletal structures associated with flight. These remnants of its avian ancestry were a testament to the evolutionary journey that brought the dodo to its flightless state. While incapable of sustained flight, the dodo's wings likely played a role in balance, display, and other non-flight-related activities.

The trade-off for flightlessness was the dodo's enhanced terrestrial abilities. With its robust build, sturdy legs, and waddling gait, the dodo was well-adapted for life on the ground. It could navigate the forests, traverse coastal regions, and forage for food with relative ease. These adaptations allowed the dodo to exploit the resources available on Mauritius, forging a niche as a ground-dwelling herbivore.

The loss of flight also led to changes in the dodo's body proportions. Its wings, no longer under the selective pressures of flight, reduced in size, while its body became more rounded and plump. This shape provided insulation and energy storage, allowing the dodo to thrive in the varying climates of its island habitat.

While flightlessness may seem like a disadvantage, the dodo's adaptation to a terrestrial lifestyle brought its own advantages. It was freed from the energy-intensive demands of flight, allowing resources to be allocated to other biological functions. The dodo's beak, for example, evolved to be robust and well-suited for cracking open fruits, nuts, and seeds—a specialized tool that aided its foraging activities.

The dodo's flightless nature also had an impact on its behavior and social dynamics. Without the ability to escape predators through flight, the dodo developed a unique approach to survival. It displayed a lack of fear towards humans and other potential threats, likely due to the absence of natural predators on Mauritius. This trusting nature ultimately contributed to its vulnerability when humans arrived on the island.

As we marvel at the flightless wonder of the dodo, we gain a deeper appreciation for the diversity of avian life and the myriad ways in which birds have adapted to their environments. The dodo's story challenges our preconceptions about flight as an essential trait, reminding us of the vast possibilities that exist within the avian world.

Adaptations for Island Life

Islands, with their unique ecological dynamics and isolated environments, have a profound influence on the species that call them home. Over time, organisms adapt to the specific challenges and opportunities presented by their island habitats. The dodo, endemic to the island of Mauritius, exemplifies the remarkable adaptations that emerge in response to the demands of island life.

One of the most significant adaptations of the dodo was its flightlessness, which allowed it to thrive in the absence of land predators. On Mauritius, where the dodo evolved, there were no natural threats that necessitated the use of flight for survival. Consequently, the dodo gradually lost its ability to fly, redirecting energy and resources towards other advantageous traits.

The loss of flight in the dodo led to changes in its anatomy. Its wings, once crucial for aerial locomotion, reduced in size and became less functional. While incapable of sustained flight, the dodo's wings retained some feathers and skeletal structures associated with flight. These remnants of its avian ancestry served other purposes, such as balance, display, and communication.

With its newfound terrestrial lifestyle, the dodo developed a robust physique. Its body became rounded and plump, providing insulation and energy storage. This adaptation allowed the dodo to navigate the varied climates of Mauritius, from the lush forests to the coastal regions, with relative ease.

The dodo's beak underwent a remarkable transformation to meet the demands of its specialized diet. Over generations, the beak evolved into a large, stout, and slightly curved structure. This shape was perfectly suited for cracking open the tough outer shells of fruits, nuts, and seeds, allowing the dodo to access the nutritious contents within. This unique beak adaptation enabled the dodo to exploit a wide range of food resources available on the island.

As a ground-dwelling bird, the dodo developed sturdy legs and feet. Its legs were relatively short and well-adapted for traversing the forest floor and sandy coastal areas. The dodo's gait, often described as a waddle, added to its distinctive charm and set it apart from other bird species.

The dodo's behavior and social dynamics were also influenced by its adaptation to island life. Without the need for flight as an escape mechanism, the dodo displayed a lack of fear towards humans and other potential threats. This trusting nature, honed in the absence of natural predators, ultimately made it vulnerable when humans arrived on Mauritius.

Furthermore, the dodo's reproductive strategies were shaped by the island's unique conditions. The bird had a relatively slow reproductive rate, laying only one or two eggs at a time. This strategy was likely an adaptation to the stable and resource-limited environment of the island.

These adaptations, both physical and behavioral, allowed the dodo to thrive in the isolated ecosystem of Mauritius. Flightlessness, a robust physique, specialized beak morphology, and altered social dynamics were all integral components of the dodo's success on the island.

The story of the dodo's adaptations for island life serves as a reminder of the incredible resilience and versatility of life forms. It showcases the remarkable ways in which organisms can transform in response to the unique challenges and opportunities presented by their environments.

The Dodo's Diet and Feeding Habits

In our exploration of the dodo, we now turn our attention to its diet and feeding habits. Like all animals, the dodo relied on sustenance to fuel its daily activities and meet its nutritional needs. Understanding what the dodo ate and how it obtained its food provides valuable insights into its ecological role and its place within the intricate web of Mauritius' ecosystem.

The dodo's diet was primarily herbivorous, consisting of a variety of plant matter. Being flightless and terrestrial, the dodo had adapted to forage on the resources available on the island of Mauritius. Fallen fruits, seeds, nuts, and the occasional plant material comprised the bulk of its diet.

One of the dodo's specialized adaptations for feeding was its beak. The dodo possessed a large, stout, and slightly curved beak that allowed it to crack open the tough outer shells of fruits, nuts, and seeds. This unique beak morphology provided the dodo with access to the nutritious contents within, enabling it to exploit a wide range of food resources.

Mauritius, with its diverse plant life, provided the dodo with ample opportunities for foraging. The island was adorned with an array of endemic plants, including ebony trees, palm species, and other fruit-bearing plants. These plants served as vital food sources for the dodo and other herbivorous creatures that inhabited the island.

The dodo's feeding habits were closely tied to the seasonal availability of food. During times of plenty, when fruits were abundant, the dodo could feast on a variety of options. Fallen fruits, such as those from the Tambalacoque tree (Sideroxylon grandiflorum), were a particular favorite. These fruits, known as dodo eggs due to their shape and size, provided a valuable source of nutrition for the bird.

In addition to fruits, the dodo would also consume seeds and nuts. By cracking open their tough shells with its specialized beak, the dodo could access the energy-rich contents within. This allowed the bird to supplement its diet during times when fruits were less available.

While the dodo's diet was predominantly herbivorous, there is some evidence to suggest that it may have occasionally consumed small animal matter. Insects, lizards, and other small animals could have been opportunistically consumed by the dodo, providing additional protein and nutrients. However, the extent and frequency of such animal consumption remain subjects of scientific inquiry.

The dodo's feeding habits were also influenced by the unique social dynamics of the species. Observations by early explorers suggest that the dodo may have exhibited communal feeding behaviors. These gatherings of dodos, often referred to as "herds" or "droves," could be seen foraging together in a coordinated manner. This behavior not only allowed them to efficiently exploit food resources but also served as a means of protection against potential threats.

Sadly, the arrival of humans on Mauritius had a profound impact on the dodo's diet and feeding habits. The dodo's trusting nature and lack of fear towards humans made it an easy target for exploitation. Sailors and settlers hunted the bird for food, disrupting its feeding patterns and contributing to its decline.

As we explore the dodo's diet and feeding habits, we gain a deeper understanding of its ecological role and its significance within the island's ecosystem. The dodo's specialized beak, its consumption of fruits, seeds, and nuts, and its communal feeding behaviors all shaped its interactions with the plants and animals of Mauritius.

The Dodo's Social Behavior

In our exploration of the dodo, we now delve into the intriguing realm of its social behavior. As social creatures, animals form complex interactions, communicate with one another, and establish social hierarchies. The dodo, with its unique place in the avian world, displayed fascinating social behaviors that shed light on its life and relationships within the island ecosystem of Mauritius.

Early accounts and observations by European explorers described the dodo as a sociable and gregarious bird. It was often encountered in groups or gatherings, displaying a level of social cohesion and coordination. These gatherings, sometimes referred to as "herds" or "droves," consisted of multiple individuals foraging and moving together.

The communal feeding behaviors of the dodo were a prominent aspect of its social dynamics. These gatherings allowed the birds to efficiently exploit food resources, sharing knowledge of feeding locations and cooperating in obtaining sustenance. By foraging together, the dodo maximized its chances of finding food and minimized the risks associated with predation.

Within these social groups, the dodo likely had a hierarchical structure, with dominant individuals exerting influence over others. While the specific details of the social hierarchy remain speculative, it is believed that factors such as size, age, and physical condition may have played a role in determining individual status. Dominant individuals may have enjoyed preferential access to resources and opportunities for mating.

Mating and courtship behaviors in the dodo were likely influenced by the social structure and hierarchy. During the breeding season, males likely engaged in displays and behaviors to attract mates and establish dominance within the group. These displays may have involved puffing up their chests, erecting their feathers, and vocalizing to impress females and intimidate rival males.

Nesting behavior in the dodo was a solitary affair. Females would seek out suitable locations to build their nests, typically on the ground. The nests were constructed using available vegetation, such as leaves, twigs, and other plant materials. Once the nest was complete, the female would lay one or two eggs and incubate them until hatching.

Parental care in the dodo was primarily the responsibility of the female. She would diligently incubate the eggs and protect the nest from potential threats. The exact duration of incubation and the level of parental investment in rearing the young remain subjects of scientific inquiry.

The social behavior of the dodo was not limited to interactions within its own species. The dodo shared its island habitat with other endemic species, such as the giant tortoise and various bird species. While the specifics of their interactions are not well-documented, it is likely that the dodo had ecological connections and dependencies with these cohabiting organisms.

The arrival of humans on Mauritius brought significant changes to the dodo's social dynamics. The bird's lack of fear towards humans made it susceptible to exploitation and hunting. The dodo's social gatherings and communal behaviors became its downfall, as groups of birds were easily targeted and decimated by sailors and settlers in search of food.

The extinction of the dodo not only marked the loss of a fascinating species but also highlighted the vulnerability of social creatures in the face of human activities. The dodo serves as a reminder of the importance of understanding and conserving social behaviors within animal populations, as they contribute to the stability and resilience of ecosystems.

Encounter with the First Humans

The arrival of humans on the remote island of Mauritius marked a turning point in the history of the dodo. For thousands of years, the dodo thrived in isolation, devoid of natural predators and relatively undisturbed by external influences. However, the encounter with the first humans would forever alter the fate of this iconic bird.

The precise details of the initial human encounter with the dodo are shrouded in the mists of time. It is believed that the first humans to set foot on the shores of Mauritius were sailors and explorers, drawn to the island's remote allure. These early visitors would have been met with a remarkable sight—the dodo, a flightless bird unlike any they had encountered before.

For the sailors and explorers, the dodo was a source of fascination and curiosity. Its trusting nature and lack of fear towards humans made it an easy target for exploitation. The dodo's communal gatherings, once an advantage for foraging, now worked against it, as groups of birds were easily hunted for their meat.

Historical records suggest that the sailors and explorers hunted the dodo for food. The dodo's plump and meaty body, combined with its lack of fear, made it a convenient source of sustenance during long voyages. The birds were captured, slaughtered, and consumed, their numbers dwindling as human appetites were satisfied.

As the dodo became more familiar to sailors, it also found its way into the accounts and sketches of early explorers.

These depictions, although often artistic interpretations, provide valuable glimpses into the physical appearance and behaviors of the dodo. The illustrations and descriptions by artists such as Roelant Savery and Francois Leguat allowed the dodo's image to permeate popular culture and ignite the public's imagination.

The impact of human arrival on the dodo extended beyond direct hunting. Humans brought with them invasive species, inadvertently introducing predators and competitors into the island ecosystem. Rats, dogs, pigs, and monkeys wreaked havoc on the dodo's habitat, consuming its eggs, raiding its nests, and competing for limited resources. Human colonization of Mauritius brought further challenges for the dodo. The clearing of forests for agriculture, the introduction of domestic animals, and habitat degradation further diminished the dodo's chances of survival. The bird's lack of fear towards humans, once a trait that served it well, became a fatal flaw as it fell prey to the expanding human presence on the island.

By the late 17th century, a mere hundred years after its encounter with humans, the dodo had vanished from the face of the Earth. The extinction of the dodo is a tragic reminder of the irreversible consequences of human activities on vulnerable species and delicate ecosystems.

The story of the dodo serves as a cautionary tale, highlighting the importance of responsible stewardship of our natural world. It reminds us of the delicate balance between human actions and the survival of unique and irreplaceable species. The dodo's fate stands as a testament to the need for conservation, preservation, and the recognition of the interdependence of all life forms.

Early Observations and Specimens

In our quest to unravel the mysteries of the dodo, we now turn our attention to the early observations and specimens that provided valuable insights into the life and characteristics of this iconic bird. The dodo, with its unique appearance and intriguing nature, captured the attention of sailors, explorers, and naturalists who encountered it during its brief existence.

The first recorded observations of the dodo were made by Dutch sailors and explorers in the late 16th century. These early encounters with the bird left a lasting impression on those who witnessed its peculiar appearance and trusting nature. Accounts of the dodo's appearance and behaviors began to circulate, piquing the interest of naturalists and sparking a desire to learn more about this enigmatic creature.

Early explorers, such as Dutchman Cornelis Matelief de Jonge, recorded their observations of the dodo during their voyages to Mauritius. These written accounts described the bird's large size, flightlessness, rounded body, and unique beak. The dodo's plump physique, combined with its curious and trusting nature, made it an easy target for capture and consumption.

In addition to written accounts, early explorers also collected physical specimens of the dodo, ranging from bones and skeletal remains to preserved skins and fragments of feathers. These specimens provided tangible evidence of the bird's existence and allowed for further study and analysis.

One notable example of early dodo specimens is the Oxford dodo, which comprises the partially preserved head and foot of a dodo. This specimen, held at the Oxford University Museum of Natural History, provides valuable insights into the bird's physical characteristics and has been a subject of scientific investigation and public fascination.

Artistic representations of the dodo also played a crucial role in documenting its appearance. Paintings and drawings by artists such as Roelant Savery and Francois Leguat depicted the dodo in vivid detail, showcasing its rounded body, large beak, and plump physique. These artistic renderings, while not always scientifically accurate, offered visual glimpses into the world of the dodo.

As the dodo became more well-known, interest in obtaining specimens grew. Sailors and settlers arriving on Mauritius collected dodo specimens for scientific study and as curiosities. These specimens, often preserved through taxidermy or skeletal preparations, were sent to various institutions and private collections around the world, becoming ambassadors for the dodo and its story.

Scientific investigations of these specimens provided crucial insights into the dodo's anatomy, physiology, and evolutionary relationships. Comparative analysis of the skeletal remains allowed scientists to better understand the dodo's closest relatives within the pigeon family, while the preserved feathers provided clues about its coloration and plumage.

Despite the efforts to preserve specimens, the passage of time took its toll. Many early dodo specimens were lost to decay, deterioration, or accidental destruction. However, a few precious specimens have survived, allowing modern scientists to continue their research and unravel the secrets of this extinct bird.

In recent years, advancements in scientific techniques have enabled further analysis of dodo specimens. DNA studies have provided valuable genetic information, shedding light on the dodo's evolutionary relationships and offering glimpses into its past. Isotopic analysis of preserved tissues has provided insights into its diet and habitat.

The early observations and specimens of the dodo provide windows into a world that existed long ago. They offer glimpses of a bird that captured the imagination of those who encountered it and sparked a desire to understand its unique biology and story. Through the careful examination and study of these early records and specimens, we continue to piece together the puzzle of the dodo's existence.

Naturalists and the Dodo's Fame

The dodo, with its unique appearance and tragic story, captured the fascination of naturalists and scientists throughout history. Its fame spread far and wide, as explorers, artists, and scholars sought to understand and document this enigmatic bird. In this chapter, we explore the contributions of naturalists to the dodo's fame and the legacy they left behind.

One of the earliest naturalists to encounter the dodo was the Dutch explorer and botanist, Pieter Hermannszoon. During his visit to Mauritius in 1602, he made detailed observations of the bird and collected specimens, contributing to our early understanding of its anatomy and behavior. Hermannszoon's accounts and sketches provided a foundation for future investigations.

As interest in the dodo grew, naturalists from various countries embarked on voyages to Mauritius to study the bird firsthand. Among them was the French naturalist, François Leguat, who spent several months on the island in the late 17th century. Leguat's detailed observations of the dodo's behavior and interactions with its environment offered valuable insights into its ecological role.

The 18th century saw increased scientific interest in the dodo. Naturalists such as Georges-Louis Leclerc, Comte de Buffon, and Johann Friedrich Gmelin studied the available dodo specimens and incorporated them into their works. Buffon, a prominent French naturalist, included a description and illustration of the dodo in his monumental work, "Histoire Naturelle," bringing the bird to the attention of a wider audience.

Artists also played a vital role in popularizing the dodo. Roelant Savery, a Flemish painter, created exquisite and imaginative depictions of the dodo based on descriptions and specimens. His paintings, filled with vivid colors and attention to detail, captured the essence of the dodo and helped shape its visual representation in the public's imagination.

The 19th century brought renewed interest in the dodo's story. Naturalists such as Richard Owen and Sir Richard Lydekker examined existing dodo specimens, analyzing their anatomy and taxonomy. Their contributions, along with those of earlier naturalists, helped establish the dodo as an iconic symbol of extinction and a cautionary tale of the consequences of human activities.

The dodo's fame extended beyond the scientific community. It permeated popular culture, inspiring literary works, artworks, and even appearing in political cartoons. Lewis Carroll referenced the dodo in his famous book "Alice's Adventures in Wonderland," further cementing its place in the public consciousness.

In the early 20th century, scientists began to question the accuracy of earlier dodo depictions and descriptions. Skeptical of the bird's rounded body and plump appearance, some proposed alternative interpretations of its anatomy and suggested that early illustrations may have exaggerated its features. These debates and discussions spurred further scientific investigation into the dodo's true physical characteristics.

Today, the dodo continues to captivate researchers and enthusiasts alike. Advances in technology have allowed for more precise reconstructions of the dodo's appearance and the study of its genetic material. DNA analysis has provided insights into its evolutionary relationships and genetic diversity, shedding light on the bird's place in the avian family tree.

The fame of the dodo owes much to the contributions of naturalists who dedicated their lives to the study of the natural world. Their meticulous observations, artistic representations, and scientific investigations ensured that the legacy of the dodo endured. Through their work, the dodo became an enduring symbol of human impact on vulnerable species and the urgent need for conservation efforts.

The Golden Age of Exploration

The Golden Age of Exploration, spanning roughly from the 15th to the 17th century, was a period of significant voyages, discoveries, and encounters that forever changed our understanding of the world. It was an era marked by bold explorers, intrepid sailors, and groundbreaking expeditions that opened up new frontiers and expanded human knowledge. In this chapter, we delve into the Golden Age of Exploration and its profound impact on our understanding of the natural world.

During this transformative period, explorers from European nations embarked on ambitious voyages of discovery, driven by a thirst for wealth, fame, and the desire to uncover new trade routes. Spain, Portugal, England, France, and the Netherlands were among the leading maritime powers of the time, each seeking to stake their claim in the race for exploration.

Advancements in shipbuilding and navigation techniques, such as the development of the caravel and the astrolabe, enabled sailors to venture farther into the unknown. These advancements, coupled with the desire for new trade routes to the East Indies, sparked a wave of exploration that would forever alter the course of history.

The search for new trade routes led explorers to set sail on perilous journeys across uncharted waters. In their quest to reach the lucrative markets of the East, they encountered diverse lands, cultures, and, of course, remarkable flora and fauna previously unknown to Europeans.

It was during this Golden Age of Exploration that the first European sailors arrived on the remote island of Mauritius, encountering the unique and extraordinary dodo. The dodo, with its flightlessness, trusting nature, and curious appearance, left a lasting impression on those early explorers, sparking curiosity and intrigue.

The encounters with the dodo became part of the larger narrative of exploration, adding to the allure of these voyages. Sailors returning to their homelands shared tales of this strange bird, captivating the imaginations of people who had never set foot on distant shores. The dodo, with its distinct features and association with a remote island, became a symbol of the exotic and the unknown.

Explorers, artists, and naturalists sought to document the wonders they encountered on their journeys. Detailed journals, sketches, and illustrations became valuable records of the plants, animals, and landscapes encountered during these expeditions. These visual and written accounts provided glimpses into the extraordinary biodiversity of the world and inspired further exploration and scientific inquiry.

The dodo, with its peculiar appearance and lack of fear towards humans, quickly became a subject of fascination. Early explorers recorded their observations of the bird's physical characteristics, behaviors, and interactions with its environment. These accounts, combined with artistic representations, added to the dodo's fame and solidified its place in the annals of natural history.

The Golden Age of Exploration not only expanded our knowledge of the natural world but also had far-reaching consequences for the human societies involved.

Exploration led to the establishment of colonial empires, the exchange of goods and ideas, and the reshaping of geopolitical landscapes. It laid the foundation for global trade networks and cultural exchange, forever altering the course of human history.

As we reflect on the Golden Age of Exploration, we recognize the courage and curiosity of those intrepid explorers who ventured into the unknown. Their discoveries not only brought new lands and peoples into the collective consciousness but also unveiled the rich tapestry of life on Earth.

Ecological Impact of Human Presence

The arrival of humans on the remote island of Mauritius had profound ecological consequences, forever altering the delicate balance of its unique ecosystem. The introduction of new species, habitat modification, and overexploitation of resources disrupted the natural dynamics that had evolved over millennia. In this chapter, we delve into the ecological impact of human presence on the island and its repercussions for the dodo and other native species.

One of the most significant ecological impacts of human presence was the introduction of invasive species. Ships visiting Mauritius unintentionally carried rats, mice, and other small mammals that found a newfound paradise devoid of natural predators. These introduced species, with their voracious appetites and ability to outcompete native fauna, wreaked havoc on the island's delicate ecosystem.

The dodo, ill-equipped to defend itself against these newfound threats, suffered greatly. The rats and other invasive species preyed on dodo eggs, raiding nests and decimating their reproductive success. With no evolutionary history of dealing with such predators, the dodo's population declined rapidly, hastening its path towards extinction.

Habitat modification also played a significant role in the ecological impact of human presence. As settlers arrived on Mauritius, they cleared vast areas of forests for agriculture, timber, and the establishment of human settlements. These actions resulted in the loss of the dodo's natural habitat and disrupted the intricate web of interactions between plants and animals.

The loss of forest cover had cascading effects on the entire ecosystem. Many native plant species, upon which the dodo and other animals relied for food and shelter, dwindled in numbers or disappeared entirely. The reduction in available food resources and habitat fragmentation further exacerbated the challenges faced by the dodo and other endemic species.

Another ecological impact was the overexploitation of resources by humans. The dodo, with its trusting nature and lack of fear towards humans, was an easy target for hunting. Sailors and settlers saw the bird as a convenient source of fresh meat, capturing and consuming large numbers of individuals. The relentless hunting pressure, combined with habitat loss, pushed the dodo closer to the brink of extinction.

The disappearance of the dodo had far-reaching consequences for the ecosystem. As a seed disperser, the dodo played a crucial role in maintaining the balance of plant species on the island. By ingesting fruits and excreting the seeds intact, the dodo contributed to the dispersal and germination of various plant species. With the loss of the dodo, the dynamics of seed dispersal were disrupted, potentially leading to changes in the composition and distribution of plant communities.

The ecological impact of human presence on Mauritius extends beyond the dodo. The introduction of domesticated animals, such as goats and pigs, had detrimental effects on native vegetation. These animals fed on native plants, trampled delicate ecosystems, and contributed to soil erosion, further degrading the habitats of native species.

The loss of the dodo and the degradation of the island's ecosystem served as a poignant reminder of the interconnectedness of species and the fragile nature of island environments. It highlighted the importance of preserving the intricate web of interactions that had evolved over millions of years, and the consequences that arise when this balance is disrupted by human activities.

Today, efforts are underway to restore and protect the remaining natural habitats of Mauritius. Conservation initiatives focus on removing invasive species, restoring native vegetation, and reintroducing threatened species. These efforts aim to not only safeguard the island's unique biodiversity but also learn from the mistakes of the past and promote sustainable practices for the future.

The ecological impact of human presence on Mauritius serves as a sobering lesson in the consequences of our actions. It underscores the urgent need for responsible stewardship of our natural world and the imperative to mitigate the negative effects of human activities on vulnerable species and delicate ecosystems.

Hunting and Predation on Dodos

The dodo, with its gentle disposition and lack of fear towards humans, fell victim to hunting and predation, contributing to its rapid decline and eventual extinction. The arrival of sailors and settlers on the island of Mauritius marked a turning point for the dodo, as it encountered predators and became an easy target for exploitation. In this chapter, we delve into the hunting and predation on dodos, shedding light on the factors that led to their tragic demise.

When sailors first encountered the dodo on Mauritius, they were struck by its trusting nature and apparent lack of fear. This made the bird an easy target for hunting, as it showed little inclination to flee from approaching humans. Sailors saw the dodo as a convenient source of fresh meat during long voyages, capturing and slaughtering the birds for consumption.

The plump and meaty body of the dodo made it an enticing target for hunters. Its large size and flightlessness further worked to its disadvantage, as the birds were relatively easy to approach and capture. Sailors and settlers exploited the dodo's vulnerability, hunting it for its meat, which provided a source of sustenance during their journeys and settlements on the island.

The communal nature of the dodo also played a role in its susceptibility to hunting. Dodos often gathered in groups or "herds" while foraging, which made it even easier for hunters to capture multiple birds in a single expedition. These gatherings, once a means of cooperative foraging, became the dodo's downfall as they facilitated mass hunting and decimated populations.

As humans settled on Mauritius, they introduced new threats to the dodo in the form of invasive species. Rats, brought inadvertently by ships, preyed on dodo eggs and young chicks, raiding nests and contributing to reproductive failure. The dodo, with no evolutionary history of dealing with such predators, was ill-equipped to defend its offspring against these new threats.

In addition to invasive rats, other introduced species such as pigs, dogs, and monkeys also posed a significant threat to the dodo's survival. These animals not only competed with the dodo for food resources but also preyed on eggs, chicks, and even adult birds. The combination of hunting by humans and predation by invasive species placed immense pressure on dodo populations, leading to their rapid decline.

The loss of habitat due to human activities further exacerbated the hunting and predation pressure on dodos. As settlers cleared vast areas of forests for agriculture and timber, they not only disrupted the dodo's natural habitat but also removed crucial protective cover. The open landscapes made it easier for predators to locate and attack dodos, further diminishing their chances of survival.

The effects of hunting and predation on dodos were not limited to direct mortality. The loss of adults meant a decrease in breeding individuals, further hindering the recovery of dodo populations. With fewer individuals available for mating, reproductive success declined, exacerbating the population decline and driving the dodo closer to extinction.

The extinction of the dodo serves as a stark reminder of the devastating consequences of unchecked hunting and the introduction of invasive species. The combination of human exploitation and predation pressure from invasive animals proved to be a lethal combination for the dodo, leading to its complete eradication from the face of the Earth.

The story of the dodo's hunting and predation highlights the need for responsible and sustainable practices in our interactions with vulnerable species and their habitats. It emphasizes the importance of conservation efforts, habitat preservation, and the prevention of invasive species introductions to protect the delicate balance of ecosystems and prevent further losses of unique and irreplaceable species.

Human Exploitation and Extinction

The story of the dodo is ultimately one of human exploitation and the tragic path towards extinction. The arrival of humans on the remote island of Mauritius had profound and irreversible consequences for this unique bird and serves as a poignant reminder of the impact our actions can have on vulnerable species. In this chapter, we delve into the human exploitation of the dodo and the factors that led to its ultimate demise.

When sailors and settlers first encountered the dodo on Mauritius, they found a bird that was trusting, curious, and lacking a natural fear of humans. This lack of fear made the dodo an easy target for exploitation. Sailors saw the bird as a convenient source of fresh meat during their long voyages, capturing and hunting it for sustenance.

The plump and meaty body of the dodo made it particularly desirable to hunters. Its large size and flightlessness further worked to its disadvantage, as the bird was relatively easy to approach and capture. The dodo became a ready source of food for sailors, who hunted it in significant numbers, leading to a rapid decline in its population.

As human settlements were established on Mauritius, the exploitation of the dodo intensified. The bird's trusting nature and communal behavior, once advantageous for cooperative foraging, now made it vulnerable to mass hunting. Dodo gatherings provided an opportunity for hunters to capture multiple birds at once, decimating populations in a short period of time.

The dodo's reproductive biology also contributed to its vulnerability. The bird had a slow rate of reproduction, with females typically laying only one or two eggs at a time. Combined with the high hunting pressure, this limited reproductive capacity meant that the dodo struggled to recover its numbers, even when hunting was not the sole cause of its decline.

In addition to direct hunting, human activities had indirect effects on the dodo's survival. The introduction of invasive species, such as rats, pigs, dogs, and monkeys, brought unintended consequences. These animals preyed on dodo eggs, chicks, and even adults, further contributing to their decline. The dodo, with no evolutionary adaptation to cope with these novel predators, was ill-equipped to defend itself or its offspring.

Habitat destruction was another significant factor in the dodo's demise. As settlers cleared vast areas of forests for agriculture and timber, they not only removed the dodo's natural habitat but also disrupted the delicate balance of the ecosystem. The loss of suitable habitats and the fragmentation of remaining areas made it increasingly challenging for the dodo to find food, seek shelter, and successfully breed.

The combination of direct hunting, predation by invasive species, and habitat degradation placed immense pressure on the dodo population. By the late 17th century, just a century after the arrival of humans on Mauritius, the dodo was gone. It had become yet another casualty of human exploitation and a testament to the devastating impact our actions can have on vulnerable species.

The extinction of the dodo serves as a stark reminder of the consequences of unchecked exploitation and the need for responsible stewardship of our natural world. It highlights the importance of conservation efforts, sustainable practices, and the recognition of the intrinsic value of all living creatures.

Today, the legacy of the dodo lives on as a symbol of the importance of preserving biodiversity and protecting vulnerable species from the threats of human activities. It serves as a reminder that we have the power and responsibility to make informed decisions and take actions that will ensure the survival of Earth's incredible and irreplaceable biodiversity.

The Last of the Dodos

The dodo, once a remarkable inhabitant of the remote island of Mauritius, met its tragic fate at the hands of human exploitation and the ecological pressures brought upon by human presence. In this chapter, we delve into the final years of the dodo and the extinction of this unique bird that forever changed the course of natural history.

By the late 17th century, the dodo's population had been decimated by hunting, predation, habitat loss, and the introduction of invasive species. The once-thriving population had dwindled to a mere handful of individuals, on the brink of extinction.

The exact details of the dodo's final years remain shrouded in uncertainty. Historical accounts suggest that the last surviving dodos were seen in the 1660s and early 1670s, as sailors and explorers reported encountering a few individuals on the island. These sightings, however, marked the end of an era for the dodo.

The last of the dodos, struggling to survive amidst the changing environment and relentless pressures, faced a grim reality. With a dwindling population and limited breeding opportunities, their chances of long-term survival were slim. The ecological balance on Mauritius had been disrupted to such an extent that recovery seemed nearly impossible.

The exact circumstances surrounding the extinction of the dodo remain a subject of speculation and debate among scientists. It is likely that a combination of factors contributed to its demise. Hunting, predation by invasive species, habitat loss, and reduced reproductive success all played significant roles in the downfall of this iconic bird.

While some reports suggest that the last surviving dodos were hunted down, others propose that the final individuals succumbed to natural causes, disease, or competition for resources. The extinction of the dodo was not a single catastrophic event but rather the culmination of centuries of human exploitation and ecological disruption.

The loss of the dodo was met with relative indifference by the general population of the time. Its extinction went largely unnoticed, with the bird fading into obscurity and becoming little more than a distant memory. It wasn't until later centuries that the significance of the dodo's disappearance became apparent.

In the 19th century, renewed interest in the dodo emerged as scientists, naturalists, and artists sought to understand and document this extinct bird. They pieced together the story of the dodo through historical accounts, sketches, preserved specimens, and fragments of remains. The legacy of the dodo became a subject of scientific inquiry and a symbol of the consequences of human impact on vulnerable species.

Today, the dodo stands as a cautionary tale and a symbol of the importance of conservation. It serves as a reminder that once a species is lost, it cannot be brought back. The extinction of the dodo was a wake-up call, sparking a global recognition of the need to protect and preserve Earth's biodiversity.

Efforts to study and learn from the dodo's story continue to this day. Scientific research, genetic analysis, and ecological studies provide valuable insights into the dodo's biology, its evolutionary relationships, and the ecological dynamics of the unique island it once called home. The dodo's legacy lives on through these ongoing investigations.

The story of the last of the dodos serves as a reminder of the responsibility we bear as stewards of our planet. It compels us to reflect on our impact on the natural world and to make informed decisions that protect and preserve the incredible diversity of life that surrounds us.

Legends and Myths Surrounding the Dodo

The dodo, with its peculiar appearance and tragic fate, has sparked a multitude of legends and myths throughout history. This chapter delves into the folklore, stories, and enduring myths that have emerged around this iconic bird, capturing the imagination of people around the world.

From the moment the dodo was first encountered by sailors and explorers, it became the subject of fascination and wonder. Tales of this unusual bird quickly spread, intertwining with local folklore and inspiring fantastical legends that have endured through the ages.

One of the most enduring myths surrounding the dodo is that it was a foolish and unintelligent creature. This misconception likely stemmed from its lack of fear towards humans, which led early observers to perceive it as an easy target for capture. However, scientific studies have since dispelled this myth, showing that the dodo was well adapted to its environment and possessed unique ecological traits.

Another myth that emerged was the idea that the dodo's meat was distasteful or poisonous. This belief likely arose due to the reports of sailors who found the dodo's flesh unpalatable after long sea voyages. However, there is no scientific evidence to support the claim that the dodo's meat was toxic or inedible. It is more likely that the taste and quality of the meat were influenced by factors such as the bird's diet or the conditions of preservation.

In popular culture, the dodo has often been portrayed as a clumsy and comical creature. This characterization can be traced back to the exaggerated depictions of the bird in artworks and literature. While the dodo's rounded body and waddling gait may have contributed to this perception, it is important to remember that these artistic representations were not always scientifically accurate.

Legends surrounding the dodo have also found their way into literature and folklore. In Lewis Carroll's "Alice's Adventures in Wonderland," the dodo is depicted as the character who organizes the "Caucus-Race," a whimsical event where everyone runs in a circle with no clear winner. The dodo's inclusion in the story has added to its enduring popularity and the perpetuation of dodo-related myths.

The extinction of the dodo also gave rise to stories of lost treasures and hidden dodo eggs. Legends circulated, suggesting that the dodo laid a single, gigantic egg that held incredible riches or possessed magical properties. These tales added an air of mystery and allure to the already mythical status of the bird.

In some cultures, the dodo has taken on symbolic meanings. It has been associated with concepts such as resilience, extinction, and the consequences of human actions. The dodo's tragic fate serves as a cautionary tale, reminding us of the fragility of life and the importance of conservation.

While legends and myths have embellished the story of the dodo, it is crucial to separate fact from fiction. The dodo was a real, albeit extinct, bird that once inhabited the island of Mauritius. Its unique characteristics and untimely demise have captivated the imaginations of people across generations, leading to the emergence of various tales and interpretations.

As we continue to study and learn about the dodo, we must separate the enduring myths from the scientific realities. By focusing on the known facts and the discoveries made through careful research, we can gain a deeper understanding of the dodo's true nature and appreciate the real wonder of this remarkable bird.

Rediscovering the Dodo in Art and Literature

The dodo, a bird long vanished from the face of the Earth, continues to captivate the imagination of artists and writers. Despite its extinction, the dodo lives on through depictions in art, literature, and popular culture. In this chapter, we explore the rediscovery of the dodo in art and literature, tracing its enduring presence in the creative realm.

From the earliest encounters with the dodo by sailors and explorers, the bird's unique appearance and story ignited the imaginations of artists. Painters, sculptors, and illustrators sought to capture the essence of this enigmatic bird, often relying on secondhand accounts and limited specimens to bring it to life on canvas and in sculptures.

One of the earliest depictions of the dodo can be found in the illustrations of Dutch artist Roelandt Savery, who was active in the early 17th century. Savery's paintings, based on his observations of captive dodos and descriptions provided by sailors, introduced the dodo to a wider audience. His detailed and vibrant renditions became influential in shaping the popular image of the bird.

As scientific interest in the dodo grew, naturalists and explorers sought to document and study the bird. They carefully sketched and described the dodo, providing valuable insights into its anatomy and behavior. These illustrations, often accompanied by detailed written accounts, served as scientific records and laid the foundation for further understanding of the species.

In addition to scientific representations, the dodo also found its way into literary works. Writers of the time were inspired by the bird's unique features and its tragic fate. The dodo became a symbol of rarity, extinction, and the consequences of human actions. References to the dodo can be found in works by renowned authors such as Jonathan Swift, Voltaire, and Lewis Carroll, further perpetuating its presence in literature.

Lewis Carroll's "Alice's Adventures in Wonderland" played a particularly significant role in popularizing the dodo. The character of the dodo, organizing the famous Caucus-Race, introduced the bird to a broader audience and forever associated it with the whimsical world of Wonderland. Carroll's inclusion of the dodo added a touch of fantasy and enchantment to the bird's already mythic status.

The dodo's presence in art and literature continued to evolve throughout the centuries. In more recent times, the dodo has become an emblem of conservation and a symbol of the urgent need to protect vulnerable species. Artists and writers use the dodo as a powerful metaphor, reminding us of the fragility of life and the importance of preserving our natural heritage.

Today, the dodo is celebrated and commemorated through various artistic mediums. Paintings, sculptures, illustrations, and even digital creations continue to portray the dodo, showcasing its distinct features and evoking a sense of wonder and nostalgia. Artistic interpretations provide a connection to a creature long gone, allowing us to appreciate its uniqueness and reflect on the consequences of human actions.

The dodo's legacy in art and literature is a testament to the enduring power of this remarkable bird. Its portrayal across different artistic mediums serves as a reminder of our fascination with the natural world and our responsibility as stewards of the Earth's biodiversity.

Scientific Analysis and Reconstruction

The dodo, long extinct and lost to the annals of time, has experienced a fascinating revival through scientific analysis and reconstruction. In this chapter, we explore the remarkable efforts undertaken by researchers to uncover the secrets of the dodo's anatomy, behavior, and ecological role, as well as their attempts to bring this extraordinary bird back to life through digital reconstructions and replicas.

The study of dodo remains, including bones, fragments, and subfossil specimens, has provided scientists with invaluable insights into the physical characteristics and biology of the bird. By carefully analyzing these remnants, researchers have been able to reconstruct the dodo's skeletal structure, estimate its size, and gain a deeper understanding of its evolutionary relationships.

The dodo's bones, particularly the large and sturdy leg bones, give us clues about its flightlessness. The bird's robust build, relatively short wings, and reduced keel bone indicate that it had lost its ability to fly, likely due to the absence of predators and the abundance of food resources on the isolated island of Mauritius.

Scientific investigations have also shed light on the dodo's diet and foraging behavior. By studying the morphology of its beak and examining plant remains found in fossilized dodo gizzards, researchers have determined that the bird was primarily herbivorous, feeding on fruits, seeds, nuts, and possibly even fallen vegetation. The dodo played a crucial role in seed dispersal on the island, contributing to the survival and distribution of various plant species.

Advancements in technology have allowed scientists to digitally reconstruct the appearance of the dodo based on the available evidence. Using computed tomography (CT) scans, 3D modeling, and comparative studies with related bird species, researchers have pieced together a more accurate representation of the dodo's external features, such as its body shape, plumage, and facial characteristics.

These digital reconstructions have challenged earlier artistic depictions of the dodo, which were often based on limited information and imaginative interpretations. The scientific reconstructions strive to present a more realistic portrayal of the bird, drawing from scientific data and a deep understanding of avian anatomy.

Replicas and models of the dodo have also been created, providing tangible representations of this extinct bird for museums, educational institutions, and public exhibitions. These replicas, carefully crafted based on scientific findings, allow people to see and interact with the dodo, fostering a deeper connection to this extraordinary creature.

Furthermore, genetic analysis has provided additional insights into the dodo's evolutionary relationships and its place within the avian family tree. By extracting and studying DNA from preserved dodo remains, scientists have been able to determine the bird's closest living relatives and better understand its genetic history.

The scientific analysis and reconstruction of the dodo continue to evolve as new technologies and research methods emerge. The ongoing quest to unlock the secrets of the dodo's biology and behavior fuels our curiosity and deepens our understanding of this unique species.

By studying the dodo, we gain valuable knowledge about the ecological dynamics of island ecosystems, the impact of human activities on vulnerable species, and the importance of conservation efforts. The scientific exploration of the dodo's legacy reminds us of the significance of preserving Earth's biodiversity and the responsibility we bear as custodians of our planet.

Clues from Fossil and Subfossil Evidence

Fossil and subfossil evidence have played a crucial role in unraveling the mysteries of the dodo's existence. These remnants of the past provide scientists with valuable clues about the dodo's physical characteristics, behavior, and the environment it once inhabited. In this chapter, we explore the fascinating insights gained from the study of fossil and subfossil evidence, shedding light on the dodo's story.

Fossils are the preserved remains or traces of organisms from past geological ages. In the case of the dodo, the available fossils are relatively limited, consisting primarily of bones, teeth, and a few fragments of eggshells. These remnants, though sparse, have provided valuable information about the dodo's anatomy and evolution.

By carefully examining the dodo's fossilized bones, scientists have been able to reconstruct the bird's skeletal structure. The size and shape of the bones have revealed that the dodo was a large, flightless bird, with a sturdy build and relatively short wings. These adaptations suggest a bird that relied more on walking and terrestrial movement rather than flight.

The study of fossilized beaks and skulls has also provided insights into the dodo's feeding habits and niche within its ecosystem. The shape and structure of the beak indicate an herbivorous diet, with a specialization for feeding on fruits, nuts, and seeds. These findings align with historical accounts and observations made by early explorers who encountered the dodo on the island of Mauritius.

Subfossil evidence, which refers to the partially fossilized remains of more recent organisms, has further enriched our understanding of the dodo. Subfossils are often found in swampy or marshy areas where the conditions for preservation are favorable. The subfossil remains of the dodo include bones, eggshell fragments, and even some soft tissues.

The preservation of soft tissues in subfossil specimens has been exceptionally rare, but in a few instances, fragments of skin, feathers, and internal organs have been discovered. These extraordinary findings have provided scientists with invaluable insights into the dodo's external appearance, such as the texture and coloration of its feathers.

Subfossils have also yielded evidence of the dodo's reproductive biology. By examining eggshell fragments, scientists have estimated the size and structure of dodo eggs. The findings suggest that the dodo laid relatively large eggs, comparable in size to those of other large flightless birds. This information gives us a glimpse into the dodo's reproductive strategy and the challenges it faced in ensuring the survival of its offspring.

The study of fossil and subfossil evidence is not without its challenges. The limited number of specimens and the fragmented nature of the remains make it difficult to piece together a complete picture of the dodo's biology and ecology. However, through careful analysis and comparison with related species, scientists have been able to draw meaningful conclusions about the dodo's evolutionary history and ecological role.

The examination of dodo fossils and subfossils has also contributed to our understanding of the broader ecosystem in which the bird lived. The analysis of plant remains found in fossilized dodo gizzards has provided insights into the dodo's diet and its role as a seed disperser. By consuming fruits and seeds, the dodo likely played a vital ecological role in the distribution and survival of various plant species on the island of Mauritius.

The exploration of fossil and subfossil evidence continues to shed new light on the dodo's story. Advances in scientific techniques, such as radiocarbon dating and DNA analysis, offer exciting opportunities to delve even deeper into the mysteries surrounding this extinct bird.

As we unravel the clues from fossil and subfossil evidence, we gain a greater appreciation for the dodo's place in Earth's natural history. These remnants of the past serve as a window into a world long gone, reminding us of the incredible diversity of life that has existed and the importance of preserving our natural heritage.

The Dodo's Place in Evolutionary History

The dodo, though now extinct, holds a significant place in the evolutionary history of Earth's biodiversity. Understanding the dodo's evolutionary relationships and its position within the avian family tree allows us to unravel its ancient origins and shed light on its unique characteristics. In this chapter, we delve into the dodo's place in evolutionary history and explore its connections to other bird species.

The dodo belonged to a group of birds known as the columbiformes, which includes pigeons and doves. Within this group, the dodo was classified in the family Raphidae, a distinct lineage of birds that inhabited the island of Mauritius. The closest living relatives of the dodo are thought to be the Nicobar pigeon (Caloenas nicobarica) and the Rodrigues solitaire (Pezophaps solitaria), both of which are also extinct.

Through comparative studies of anatomical features, genetic analysis, and examination of shared characteristics, scientists have been able to piece together the dodo's evolutionary relationships. These studies suggest that the dodo, the Nicobar pigeon, and the Rodrigues solitaire shared a common ancestor and branched off into separate lineages over millions of years.

The dodo's flightlessness was likely an adaptation that evolved independently from its closest relatives. It is believed that the absence of predators on the island of Mauritius provided the evolutionary pressure for the dodo to lose its ability to fly. Over time, the dodo's wings became smaller, its keel bone (which provides support for flight muscles) regressed, and its body became larger and more robust.

Despite its flightless state, the dodo retained some ancestral features found in its columbiform relatives. Its beak structure, for example, bears similarities to that of other pigeons and doves. The dodo's beak was adapted for herbivorous feeding, particularly for consuming fruits, seeds, and nuts.

The dodo's place in evolutionary history offers insights into the processes of speciation, adaptation, and extinction. It serves as a remarkable example of how unique island environments can shape the evolution of species. The isolation of Mauritius allowed for the evolution of distinct forms, including the dodo, as well as the Rodrigues solitaire and other endemic species.

The dodo's evolutionary lineage, characterized by gigantism, flightlessness, and unique ecological traits, showcases the incredible diversity that can arise through adaptive radiation. In the absence of significant competition and predation, the dodo and its relatives filled ecological niches that would have been occupied by other bird species in different environments.

Studying the dodo's place in evolutionary history also provides valuable insights into the dynamics of extinction. The loss of the dodo and its relatives highlights the vulnerability of isolated island species to human-induced changes and the consequences of habitat destruction, invasive species, and overhunting. The story of the dodo serves as a cautionary tale, reminding us of the delicate balance of ecosystems and the importance of preserving biodiversity.

As scientific research continues to advance, our understanding of the dodo's place in evolutionary history may further evolve. New techniques in genetic analysis, comparative anatomy, and computational modeling offer exciting avenues for deeper insights into the dodo's lineage and its connections to other bird species.

Conservation Efforts and Lessons Learned

The story of the dodo is not only a tale of extinction but also a powerful reminder of the importance of conservation efforts and the lessons we can learn from our interactions with the natural world. In this chapter, we explore the ongoing conservation initiatives inspired by the dodo and the valuable lessons we have gleaned from its tragic fate.

The dodo's extinction served as a wake-up call, prompting a global recognition of the need to protect and conserve vulnerable species and their habitats. It sparked a shift in public awareness and a growing realization of the consequences of human actions on Earth's biodiversity. The dodo became an emblem of the consequences of unchecked exploitation, habitat loss, and the introduction of invasive species.

Conservation efforts in Mauritius, the dodo's native land, have been instrumental in protecting and preserving the island's remaining endemic species. Recognizing the importance of safeguarding its unique biodiversity, the Mauritian government has established national parks, nature reserves, and stringent conservation measures to ensure the survival of endangered flora and fauna.

One remarkable example of conservation success in Mauritius is the recovery of the Mauritius kestrel (Falco punctatus), another critically endangered bird species. Through intensive conservation efforts, including captive breeding programs, habitat restoration, and the control of invasive species, the Mauritius kestrel population has rebounded from the brink of extinction.

The lessons learned from the dodo's demise extend far beyond the shores of Mauritius. The dodo's extinction serves as a poignant reminder of the interconnectedness of species and ecosystems and the profound impact that human activities can have on fragile environments. It has highlighted the importance of taking a holistic approach to conservation that considers not only individual species but also the larger ecological context.

The dodo's story has underscored the significance of early intervention and proactive conservation measures. By recognizing the warning signs and taking action before species populations reach critically low levels, we can increase the likelihood of successful recovery and prevent irreparable loss. This includes protecting habitats, controlling invasive species, and implementing sustainable practices that minimize our impact on vulnerable ecosystems.

Education and raising public awareness have also played vital roles in conservation efforts inspired by the dodo. The dodo's tale has been shared through books, documentaries, and exhibits, helping to foster a sense of empathy and responsibility for the natural world. By educating future generations about the importance of biodiversity and the consequences of human actions, we can cultivate a deeper appreciation for the need to protect and conserve our planet's remarkable creatures.

The dodo's story also highlights the significance of collaborative conservation initiatives. Scientists, conservationists, governments, local communities, and international organizations have joined forces to develop and implement conservation strategies. Through partnerships and knowledge sharing, we can pool resources, expertise, and innovative solutions to address the complex challenges facing biodiversity conservation.

The dodo's legacy inspires us to embrace a more sustainable approach to our interactions with the natural world. It reminds us of the need to balance human needs with the preservation of ecosystems and the species that depend on them. By adopting practices such as sustainable agriculture, responsible tourism, and the reduction of our ecological footprint, we can minimize our impact and promote the long-term health of our planet.

Ultimately, the dodo's tragic extinction serves as a call to action. It compels us to reflect on our role as stewards of the Earth and to make conscious choices that prioritize the well-being of our planet's biodiversity. The lessons we have learned from the dodo's story continue to guide conservation efforts and shape our understanding of the interconnectedness of all living things.

Modern-Day Representations of the Dodo

The dodo, a bird long extinct, continues to captivate our imagination in the modern world. Its unique appearance, tragic story, and symbolic significance have made the dodo a popular subject in various forms of media, art, and popular culture. In this chapter, we explore the modern-day representations of the dodo and its enduring presence in our society.

One of the most prevalent mediums through which the dodo is represented today is in literature. Numerous books, both fiction and non-fiction, have featured the dodo as a central or supporting character. Authors draw upon the dodo's intriguing history to weave tales that entertain and educate readers. These literary works range from historical accounts and scientific studies to fantastical stories that imagine a world where the dodo still exists.

In visual arts, the dodo remains a popular subject for painters, illustrators, and sculptors. Artists strive to capture the essence of the bird, often drawing inspiration from historical accounts, scientific reconstructions, and their own imagination. These artistic representations range from realistic portrayals to more stylized interpretations, each contributing to the ever-evolving visual legacy of the dodo.

The dodo's image is also frequently employed in advertising and branding. Its distinctive appearance and recognition as a symbol of extinction and environmental conservation make it a powerful icon for various causes. The dodo's likeness can be found on logos, merchandise, and promotional materials, serving as a reminder of the need to protect endangered species and their habitats.

In the field of animation and film, the dodo has made appearances in both live-action and animated productions. Whether as a character in animated movies or as a symbol of a bygone era in historical films, the dodo's presence on the screen continues to capture the hearts of audiences worldwide. These representations often introduce the dodo to new generations and contribute to its enduring popularity.

The dodo's significance is not limited to the arts and entertainment industry. It is also celebrated in scientific circles, where its study continues to yield new discoveries and insights. Researchers utilize advanced technologies, such as DNA analysis and computer modeling, to delve deeper into the dodo's biology, behavior, and evolutionary history. These scientific representations strive to enhance our understanding of the bird and its place in the natural world.

In addition to traditional media, the dodo has found a prominent place in digital platforms and virtual spaces. Online communities, social media, and video games have embraced the dodo as an emblem of uniqueness and curiosity. Virtual representations of the dodo in video games and virtual reality experiences allow users to interact with the bird, fostering a deeper connection to its story and promoting awareness of its ecological significance.

The dodo's popularity in modern-day representations extends beyond its scientific and artistic realms. It has become a cultural symbol, often associated with ideas of resilience, extinction, and the consequences of human actions. The dodo's story serves as a reminder of the fragility of life and the importance of preserving our planet's biodiversity.

Through modern-day representations, the dodo continues to leave its mark on our collective consciousness. It serves as a constant reminder of the profound impact of extinction and the need for conservation efforts. The dodo's unique characteristics, tragic fate, and enduring symbolism make it a powerful ambassador for the protection of endangered species and the preservation of Earth's natural heritage.

As we conclude our journey through the dodo's story, we reflect on the diverse representations of this remarkable bird in the modern world. The dodo's image persists in literature, art, advertising, film, science, and digital media, keeping its legacy alive and inspiring us to appreciate the beauty and wonder of our planet's extraordinary creatures.

The Legacy of the Dodo: Lessons for Future Generations

The story of the dodo is not merely a tale of a bird lost to extinction; it is a powerful legacy that carries valuable lessons for future generations. As we conclude our exploration of the dodo's remarkable journey, we reflect on the enduring legacy it has left behind and the insights it offers for our stewardship of Earth's biodiversity.

First and foremost, the dodo serves as a stark reminder of the irreversible consequences of human actions on the natural world. The dodo's extinction was directly linked to human activities, including habitat destruction, hunting, and the introduction of invasive species. Its fate illustrates the need for responsible and sustainable interactions with our environment, highlighting the impact that our actions can have on vulnerable species.

The dodo's tragic demise has ignited a global awareness of the importance of conservation. It serves as a symbol of the urgent need to protect and preserve Earth's rich tapestry of life. The dodo's story has inspired numerous conservation initiatives and efforts to safeguard endangered species and their habitats. It teaches us that proactive conservation measures, education, and public awareness are crucial in ensuring the survival of vulnerable species.

Another valuable lesson we learn from the dodo is the significance of maintaining the delicate balance of ecosystems. The dodo's extinction disrupted the intricate web of interactions within its island habitat, causing ripple effects on other species and the overall ecosystem. It underscores the interconnectedness of life and the importance of preserving biodiversity to maintain the health and resilience of our planet.

The dodo's unique ecological role as a seed disperser highlights the importance of keystone species and their contributions to ecosystem functioning. By consuming fruits and spreading seeds, the dodo played a crucial role in maintaining the diversity and composition of plant communities. Its loss not only impacted the bird itself but also had cascading effects on the vegetation and other organisms that relied on its services. This teaches us the value of preserving not only individual species but also the ecological processes they facilitate.

Furthermore, the dodo's story teaches us the value of scientific inquiry and the importance of learning from the past. Through the study of fossils, subfossils, and historical accounts, scientists have been able to piece together the dodo's biology, behavior, and evolutionary history. The scientific legacy of the dodo highlights the power of observation, analysis, and collaboration in unraveling the mysteries of the natural world. It serves as an inspiration for future generations of scientists to continue exploring and studying Earth's biodiversity.

The dodo also imparts a sense of wonder and curiosity about the diversity of life that once graced our planet. It reminds us of the extraordinary creatures that have come before us and the immense responsibility we bear in ensuring the survival of those that remain. The dodo's unique appearance and tragic fate make it an enduring symbol of our planet's remarkable and fragile natural heritage.

Finally, the dodo's legacy encourages us to strive for a future in which we coexist harmoniously with nature. It challenges us to reassess our values, lifestyles, and consumption patterns, seeking sustainable solutions that minimize our impact on the environment. The dodo's story serves as a call to action, reminding us that we have the power to shape a world where the diversity of life can flourish.

As we conclude our book, we carry with us the lessons of the dodo's legacy. Its story urges us to become ambassadors for conservation, to champion the preservation of Earth's biodiversity, and to cultivate a deep respect for the interconnectedness of all living beings. May the legacy of the dodo serve as a compass for future generations, guiding us toward a future where the wonder and beauty of our natural world can thrive.

Printed in Great Britain
by Amazon